CHAPTER LIST:

BOOK INTRODUCTION

In the age of rapid technological advancement, few innovations have captured the imagination and potential for transformation quite like Artificial Intelligence (AI) and Machine Learning (ML). As the world witnesses an unprecedented integration of these technologies into various aspects of daily life, India stands at the forefront of this revolution. "Revolutionizing India: The Impact of Artificial Intelligence & Machine Learning" delves into the multifaceted implications of AI and ML within the Indian context.

With a population exceeding a billion and a burgeoning tech-savvy workforce, India is poised to leverage AI and ML to address longstanding challenges while propelling itself into a future defined by innovation and efficiency. From healthcare to agriculture, finance to education, the applications of AI and ML are reshaping industries and redefining possibilities. However, with these opportunities come significant considerations regarding ethics, privacy, and inclusivity, which this book explores in depth.

Through a meticulous examination of case studies, expert insights, and policy analysis, this book offers a comprehensive overview of the impact of AI and ML in India. From government initiatives to grassroots innovation, from urban development to rural empowerment, the narrative navigates through the complexities and nuances of India's AI and ML landscape. Moreover, it envisions a future where India emerges not only as a consumer but as a global leader in AI and ML technologies.

CHAPTER 1

THE GENESIS OF AI AND ML IN INDIA

India's tryst with Artificial Intelligence and Machine Learning dates back to the early days of computer science research. While the term "Artificial Intelligence" was coined in 1956, it wasn't until the late 20th century that India began making significant strides in this field. The establishment of research institutions and academic programs laid the foundation for India's journey into AI and ML.

One pivotal moment came in the 1980s with the founding of the Centre for Artificial Intelligence and Robotics (CAIR) by the Defense Research and Development Organization (DRDO). CAIR played a crucial role in pioneering AI applications for defense and strategic purposes. Concurrently, academic institutions such as the Indian Institutes of Technology (IITs) and the Indian Institutes of Science (IISc) began offering courses and conducting research in AI and ML.

Throughout the 1990s and early 2000s, India witnessed a burgeoning interest in software development and IT services, laying the groundwork for AI and ML adoption across industries. The emergence of startups and technology hubs further accelerated this trend, with companies exploring AI-driven solutions for diverse challenges.

However, it wasn't until the 2010s that AI and ML gained widespread attention in India, fueled by advances in computing power, data availability, and algorithmic innovations. Multinational corporations and homegrown startups alike began investing in AI research and development, recognizing its potential to revolutionize business processes and customer experiences.

Today, India stands as a dynamic ecosystem for AI and ML innovation, characterized by a vibrant startup culture, collaborative research initiatives, and ambitious government policies. From healthcare diagnostics to personalized recommendations, from predictive analytics to autonomous vehicles, the applications of AI and ML continue to expand, shaping India's trajectory towards a technologically empowered future.

CHAPTER 2

PIONEERING TECHNOLOGIES: AI AND ML APPLICATIONS ACROSS INDUSTRIES

The advent of Artificial Intelligence (AI) and Machine Learning (ML) has unleashed a wave of transformative innovation across industries, revolutionizing the way businesses operate and interact with customers. From healthcare to finance, from agriculture to manufacturing, the applications of AI and ML are as diverse as they are profound, promising unprecedented efficiency, accuracy, and insights. In this chapter, we explore some pioneering applications of these technologies across various sectors in India.

HEALTHCARE:

In the healthcare sector, AI and ML are revolutionizing patient care, diagnosis, and treatment. AI-powered medical imaging systems are enhancing the accuracy of diagnostics, enabling early detection of diseases such as cancer and tuberculosis. Machine Learning algorithms are being deployed to analyze vast amounts of genomic data, paving the way for personalized medicine and targeted therapies. Furthermore, AI-driven chatbots and virtual assistants are streamlining patient engagement and support, providing round-the-clock access to medical advice and information.

FINANCE:

In the finance industry, AI and ML are reshaping the landscape of banking, insurance, and investment. Fraud detection algorithms leverage machine learning to detect suspicious patterns and anomalies in transaction data, safeguarding financial institutions and their customers from fraudulent activities. AI-powered robo-advisors offer personalized investment advice and portfolio management services, democratizing access to wealth management tools. Moreover, Natural Language Processing (NLP) algorithms are transforming

customer service and support, enabling banks and insurance companies to automate routine inquiries and streamline
operations.

AGRICULTURE:

In agriculture, AI and ML technologies hold the promise of addressing food security challenges and optimizing resource utilization. Precision agriculture techniques, powered by AI and ML, leverage data from satellites, drones, and IoT sensors to monitor crop health, predict yields, and optimize irrigation and fertilization
schedules. Machine learning algorithms analyze weather patterns, soil composition, and historical data to provide actionable insights to farmers, enabling them to make informed decisions and maximize crop productivity. Additionally, AI-driven pest detection systems help farmers identify and mitigate pest infestations early, reducing crop losses and increasing yields.

MANUFACTURING:

In the manufacturing sector, AI and ML are driving efficiency, productivity, and quality assurance across the production process. Predictive maintenance algorithms leverage machine learning to analyze equipment sensor data and predict potential failures before they occur, minimizing downtime and maintenance costs. AI-powered robotic systems automate repetitive tasks such as assembly, packaging, and quality control, improving throughput and reducing errors. Furthermore, ML algorithms optimize supply chain management, forecasting demand, and identifying cost-saving opportunities.

RETAIL:

In retail, AI and ML technologies are revolutionizing customer engagement, sales forecasting, and inventory management. Recommendation systems, powered by machine learning algorithms, analyze customer behavior and preferences to deliver personalized product recommendations, driving sales and customer satisfaction.

Predictive analytics algorithms forecast demand and optimize inventory levels, minimizing stockouts and overstock situations. Moreover, AI-driven chatbots and virtual assistants provide personalized customer support and assistance, enhancing the overall shopping experience.

In conclusion, the applications of Artificial Intelligence and Machine Learning across industries in India are vast and varied, promising to reshape business models, enhance operational efficiency, and improve customer experiences. As organizations continue to embrace these technologies and harness their potential, the future holds boundless opportunities for innovation and growth.

CHAPTER 3

TRANSFORMING EDUCATION: AI AND ML IN LEARNING SYSTEMS

The realm of education is undergoing a profound transformation fueled by the integration of Artificial Intelligence (AI) and Machine Learning (ML) technologies. These advancements are revolutionizing traditional learning systems, offering personalized experiences, and unlocking new avenues for knowledge dissemination. In this chapter, we delve into the ways AI and ML are reshaping education in India.

PERSONALIZED LEARNING:

One of the most significant impacts of AI and ML in education is the ability to provide personalized learning experiences tailored to individual student needs. Adaptive learning platforms leverage ML algorithms to analyze student performance data and adapt the curriculum in real-time, providing targeted interventions and customized learning paths. This approach ensures that students receive the support and resources they need to succeed, regardless of their learning pace or style.

INTELLIGENT TUTORING SYSTEMS:

AI-powered intelligent tutoring systems are revolutionizing the way students engage with course materials and receive feedback. These systems use natural language processing and machine learning algorithms to interact with students in a

conversational manner, answering questions, providing explanations, and offering personalized recommendations. By simulating the role of a human tutor, these

systems enhance student engagement and comprehension, fostering deeper learning experiences.

AUTOMATED ASSESSMENT:

AI and ML technologies are streamlining the assessment process, enabling educators to provide timely and accurate feedback to students. Automated grading systems leverage machine learning algorithms to evaluate student assignments, quizzes, and exams, reducing the administrative burden on teachers and enabling them to focus on instructional activities. Moreover, these systems can provide detailed analytics on student performance, identifying areas of strength and weakness and informing instructional decision-making.

PREDICTIVE ANALYTICS:

In addition to supporting student learning, AI and ML are also being used to improve institutional effectiveness and student outcomes. Predictive analytics models analyze vast amounts of data, including demographic information, academic performance, and engagement metrics, to identify at-risk students and intervene proactively. By
flagging students who may be struggling academically or socially, educators can provide targeted support and interventions, ultimately improving retention and graduation rates.

CONTENT CREATION AND CURATION:

AI and ML technologies are also playing a crucial role in content creation and curation, enabling educators to develop engaging and interactive learning materials. Natural language generation algorithms can generate text-based content such as quizzes, study guides, and interactive lessons, saving educators time and effort.

Additionally, recommendation systems can suggest relevant resources and materials based on student preferences and learning objectives, enriching the learning experience and promoting self-directed learning.

In conclusion, the integration of AI and ML technologies into learning systems has the potential to revolutionize education in India, offering personalized experiences, improving student outcomes, and empowering educators. As these technologies continue to evolve and mature, they will play an increasingly central role in shaping the future of education, unlocking new opportunities for innovation and growth.

CHAPTER 4

THE ROLE OF AI AND ML IN HEALTHCARE ADVANCEMENT

Artificial Intelligence (AI) and Machine Learning (ML) are revolutionizing the healthcare industry, offering transformative solutions to longstanding challenges and driving advancements in patient care, diagnosis, and treatment. In this chapter, we explore the profound impact of AI and ML on healthcare advancement in India.

MEDICAL IMAGING AND DIAGNOSTICS:

AI and ML technologies are revolutionizing medical imaging and diagnostics, enhancing the accuracy and efficiency of disease detection. Machine learning algorithms analyze complex medical images such as X-rays, MRIs, and CT scans, identifying subtle patterns and anomalies that may indicate the presence of diseases such as cancer, tuberculosis, and cardiovascular disorders. By augmenting the capabilities of radiologists and pathologists, these technologies enable earlier and more accurate diagnoses, leading to improved patient outcomes and survival rates.

PREDICTIVE ANALYTICS AND RISK STRATIFICATION:

AI and ML algorithms are also being used to predict disease risk and stratify patients based on their likelihood of developing certain conditions. By analyzing large datasets containing clinical, genetic, and lifestyle information, predictive analytics models can identify individuals at high risk of developing chronic diseases such as diabetes,

hypertension, and cardiovascular disease. This enables healthcare providers to implement targeted interventions and preventive measures, such as lifestyle modifications, early screenings, and personalized treatment plans, to mitigate risk and improve patient outcomes.

DRUG DISCOVERY AND DEVELOPMENT:

AI and ML are transforming the drug discovery and development process, accelerating the identification and validation of novel therapeutic targets and compounds. Machine learning algorithms analyze vast repositories of biological and chemical data, predicting the efficacy and safety of potential drug candidates and optimizing drug design parameters. By streamlining the drug discovery pipeline, these technologies enable pharmaceutical companies to bring new treatments to market faster and more cost-effectively, addressing unmet medical needs and improving patient access to innovative therapies.

CLINICAL DECISION SUPPORT SYSTEMS:

AI-powered clinical decision support systems (CDSS) are empowering healthcare providers with real-time insights and recommendations at the point of care. These systems leverage machine learning algorithms to analyze patient data, including electronic health records (EHRs), medical imaging results, and genomic information, providing clinicians with evidence-based guidelines, treatment options, and risk assessments. By assisting clinicians in making more informed and personalized decisions, CDSS can improve diagnostic accuracy, treatment efficacy, and patient safety, ultimately enhancing the quality of care delivered to patients.

REMOTE MONITORING AND TELEMEDICINE:

In the era of digital health, AI and ML technologies are enabling remote monitoring and telemedicine solutions, expanding access to healthcare services and improving patient engagement and adherence. Wearable devices equipped with AI algorithms can continuously monitor vital signs, detect early signs of deterioration, and alert healthcare providers to potential emergencies. Telemedicine platforms powered by machine learning can facilitate virtual consultations, remote diagnosis, and medication management, enabling patients to receive timely and convenient care from the comfort of their homes. This is particularly relevant in India, where access to healthcare services in rural and underserved areas is limited, and telemedicine can bridge the gap and improve health outcomes for millions of people.

In conclusion, the integration of AI and ML technologies into healthcare is driving unprecedented advancements in patient care, diagnosis, and treatment in India. By harnessing the power of data-driven insights, predictive analytics, and intelligent decision support, these technologies have the potential to revolutionize the healthcare landscape, making it more personalized, efficient, and accessible for all. As AI and ML continue to evolve and mature, their impact on healthcare advancement will only grow, ushering in a new era of precision medicine and transformative innovation.

CHAPTER 5

AI, ML, AND AGRICULTURE: CULTIVATING INNOVATION

In the agricultural sector, the integration of Artificial Intelligence (AI) and Machine Learning (ML) technologies is fostering a new era of innovation and efficiency, revolutionizing traditional farming practices and addressing pressing challenges such as food security, sustainability, and climate change. In this chapter, we explore how AI and ML are cultivating innovation in agriculture in India.

PRECISION AGRICULTURE:

Precision agriculture techniques powered by AI and ML are enabling farmers to optimize resource utilization and maximize crop yields. By leveraging data from satellites, drones, IoT sensors, and weather stations, these technologies provide farmers with real-time insights into soil moisture levels, crop health, and pest infestations. Machine learning algorithms analyze this data to generate actionable recommendations for precise irrigation scheduling, targeted pesticide application, and optimal planting strategies, minimizing input costs and environmental impact while maximizing productivity.

CROP MONITORING AND DISEASE DETECTION:

AI and ML algorithms are revolutionizing crop monitoring and disease detection, enabling early intervention and mitigation of crop losses. Remote sensing technologies combined with machine learning can analyze satellite imagery and aerial drone footage to identify signs of stress, nutrient deficiencies, and disease outbreaks in crops. By detecting these issues at an early stage, farmers can take timely corrective measures, such as adjusting fertilizer applications or implementing integrated pest management strategies, to prevent yield losses and ensure crop health.

PREDICTIVE ANALYTICS FOR YIELD FORECASTING:

Predictive analytics models powered by AI and ML are transforming yield forecasting and production planning in agriculture. By analyzing historical data on weather patterns, soil conditions, and crop performance, these models can predict future
yields with a high degree of accuracy. This enables farmers to make informed decisions regarding crop selection, planting schedules, and marketing strategies, optimizing profitability and risk management. Moreover, by anticipating potential
yield fluctuations, farmers can better prepare for market dynamics and supply chain disruptions, ensuring food security and stability in the face of uncertainty.

SMART FARMING AND AUTONOMOUS MACHINERY:

AI-driven smart farming solutions and autonomous machinery are revolutionizing farm operations, reducing labor requirements and increasing efficiency. IoT

sensors
and AI algorithms enable real-time
monitoring and control of agricultural
machinery and equipment, optimizing
operations such as irrigation, fertilization,
and harvesting.
Autonomous vehicles equipped with AI-
powered navigation systems can perform
tasks such as planting, weeding, and spraying
with precision and efficiency, freeing up
farmers' time and resources for other
activities. This automation not only increases
productivity but also reduces the
environmental impact of farming by
minimizing inputs and emissions.

MARKET FORECASTING AND PRICE OPTIMIZATION:

AI and ML technologies are also being used
to analyze market trends and optimize
pricing strategies for agricultural products.
Predictive analytics models can analyze data
from various sources, including commodity
prices, weather forecasts, and consumer
demand patterns, to forecast market
conditions and price fluctuations. This

enables farmers to make strategic decisions regarding crop selection, harvest timing, and marketing channels, maximizing profitability and competitiveness in the marketplace. Moreover, by optimizing supply chain logistics and distribution networks, AI-powered solutions can reduce post-harvest losses and ensure timely delivery of fresh produce to consumers.

In conclusion, the integration of AI and ML technologies into agriculture is catalyzing innovation and transformation, empowering farmers to overcome challenges and unlock new opportunities for sustainable growth and prosperity. By harnessing the power of data-driven insights, predictive analytics, and intelligent automation, these technologies have the potential to revolutionize the way food is produced, distributed, and consumed in India, ensuring a brighter and more resilient future for the agricultural sector.

CHAPTER 6

AI AND ML IN FINANCE: SHAPING THE FUTURE OF BANKING AND INVESTMENTS

In the financial industry, the integration of Artificial Intelligence (AI) and Machine Learning (ML) technologies is reshaping traditional banking and investment practices, driving innovation, and enhancing efficiency. From personalized customer experiences to algorithmic trading, AI and ML are revolutionizing the way financial institutions operate and deliver value to their clients. In this chapter, we explore how AI and ML are shaping the future of banking and investments in India.

PERSONALIZED BANKING EXPERIENCES:

AI and ML technologies are enabling financial institutions to deliver personalized banking experiences tailored to individual

customer needs and preferences.
Customer-facing AI chatbots and virtual
assistants provide 24/7 support, answering
inquiries, resolving issues, and guiding
customers through various banking processes.
Machine learning algorithms analyze customer
transaction data and behavior
patterns to offer personalized product
recommendations, such as credit cards, loans,
and investment opportunities, enhancing
customer engagement and satisfaction.

FRAUD DETECTION AND RISK MANAGEMENT:

AI and ML algorithms are revolutionizing
fraud detection and risk management in the
financial industry, helping institutions
safeguard against fraudulent activities and
mitigate risk exposure. Machine learning
models analyze transactional data in real-
time, identifying suspicious patterns and
anomalies indicative of fraudulent behavior,
such as unauthorized transactions or identity
theft. By flagging potential fraud cases for
further investigation, these technologies
enable banks and financial institutions to
prevent losses and protect their customers'

assets and identities.

ALGORITHMIC TRADING AND PORTFOLIO Management:

AI and ML technologies are transforming investment strategies and portfolio management through algorithmic trading and predictive analytics. Machine learning algorithms analyze vast amounts of financial data, including market trends, historical prices, and news sentiment, to identify profitable trading opportunities and optimize investment portfolios. Automated trading algorithms execute trades at lightning speed, leveraging complex trading strategies and risk management techniques to maximize returns while minimizing volatility and downside risk.

CREDIT SCORING AND LENDING DECISIONS:

AI and ML algorithms are revolutionizing credit scoring and lending decisions, enabling financial institutions to assess creditworthiness more accurately and efficiently. Machine learning models analyze a wide range of data sources, including credit reports, income statements, and behavioral data, to predict the likelihood of loan repayment and assess default risk. By leveraging alternative data sources and advanced predictive analytics techniques, these technologies enable banks to extend credit to underserved populations and make more informed lending decisions, fostering financial inclusion and economic empowerment.

REGULATORY COMPLIANCE AND ANTI-MONEY LAUNDERING (AML):

AI and ML technologies are assisting financial institutions in meeting regulatory compliance requirements and combating financial crime, such as money laundering and terrorist financing. Machine learning algorithms analyze vast datasets to detect suspicious activities and transactions, flagging potential compliance violations for further investigation. By automating routine compliance tasks and enhancing detection capabilities, these technologies enable banks to streamline regulatory reporting processes, reduce compliance costs, and enhance the effectiveness of AML and Know Your Customer (KYC) programs.

In conclusion, the integration of AI and ML technologies into banking and investments is shaping the future of the financial industry in India, driving innovation, enhancing efficiency, and improving customer experiences. By harnessing the power of data-driven insights, predictive analytics, and intelligent automation, financial institutions can unlock new opportunities for growth and competitiveness in an increasingly digital and dynamic marketplace.

CHAPTER 7

SMART CITIES: URBAN DEVELOPMENT WITH AI AND ML INTEGRATION

The concept of smart cities represents a paradigm shift in urban development, leveraging technology to enhance efficiency, sustainability, and quality of life for residents. Artificial Intelligence (AI) and Machine Learning (ML) are playing a pivotal role in this transformation, enabling cities to optimize infrastructure, enhance public services, and improve governance. In this chapter, we explore how AI and ML integration are shaping the development of smart cities in India.

INTELLIGENT INFRASTRUCTURE MANAGEMENT:

AI and ML technologies are revolutionizing the management of urban infrastructure, optimizing resource allocation and maintenance schedules to ensure optimal performance and longevity. Smart sensors embedded in critical infrastructure, such as roads, bridges, and utilities, continuously collect data on usage patterns, environmental conditions, and structural integrity. Machine learning algorithms analyze this data to predict potential failures, prioritize maintenance tasks, and optimize infrastructure investments, reducing downtime and enhancing resilience against natural disasters and other disruptions.

TRAFFIC MANAGEMENT AND OPTIMIZATION:

AI and ML algorithms are transforming traffic management and optimization in urban areas, alleviating congestion and improving mobility for residents and commuters. Smart traffic monitoring systems equipped with sensors and cameras collect real- time data on traffic flow, congestion levels, and accident hotspots. Machine learning models analyze this data to predict traffic patterns, optimize signal timings, and suggest alternative routes, reducing travel

times and emissions while improving safety
and reliability.

PUBLIC SAFETY AND SECURITY:

AI and ML technologies are enhancing public
safety and security in smart cities,
enabling law enforcement agencies to
prevent and respond to crime more
effectively. Video surveillance systems
equipped with AI-powered analytics can
detect suspicious behavior, identify wanted
individuals, and alert authorities to potential
threats in
real-time. Machine learning algorithms
analyze historical crime data to predict crime
hotspots and allocate resources proactively,
deterring criminal activity and improving
overall safety perceptions among residents.

ENERGY MANAGEMENT AND SUSTAINABILITY:

AI and ML algorithms are driving energy
management and sustainability initiatives in
smart cities, optimizing energy consumption
and reducing environmental impact.

Smart grids equipped with AI-powered analytics can balance supply and demand in real-time, optimizing energy generation, distribution, and storage to minimize wastage and maximize efficiency. Machine learning models analyze data from sensors and smart meters to predict energy usage patterns, identify opportunities for energy conservation, and optimize renewable energy integration, reducing carbon emissions and promoting sustainability.

CITIZEN ENGAGEMENT AND GOVERNANCE:

AI and ML technologies are fostering citizen engagement and participatory governance in smart cities, empowering residents to actively contribute to decision-making processes and shape their communities. Smart city platforms equipped with AI-powered chatbots and virtual assistants provide residents with access to information and services, enabling them to report issues, provide feedback, and access government resources and programs. Machine learning algorithms analyze citizen feedback and sentiment data to identify emerging trends and prioritize initiatives, fostering transparency, accountability, and responsiveness in city governance.

In conclusion, the integration of AI and ML technologies into urban development is transforming the concept of smart cities in India, driving innovation, sustainability, and inclusivity. By harnessing the power of data-driven insights, predictive analytics, and intelligent automation, smart cities can optimize resources, enhance services, and improve quality of life for residents, ensuring a brighter and more resilient future for urban communities.

CHAPTER 8

CHALLENGES AND OPPORTUNITIES: ETHICAL CONSIDERATIONS IN AI AND ML

As Artificial Intelligence (AI) and Machine Learning (ML) technologies continue to advance rapidly, they bring forth a myriad of challenges and opportunities, particularly in the realm of ethics. In this chapter, we delve into the ethical considerations surrounding the development, deployment, and impact of AI and ML in various domains.

DATA PRIVACY AND SECURITY:

One of the foremost ethical considerations in AI and ML revolves around data privacy and security. As these technologies rely heavily on vast amounts of data to train models and make predictions, ensuring the privacy and security of sensitive information becomes paramount. Concerns arise regarding data breaches, unauthorized access, and the potential misuse of personal data for discriminatory or malicious purposes. Addressing these challenges requires robust data governance frameworks, transparent data practices, and stringent security measures to safeguard individuals' privacy rights and mitigate risks associated with data exploitation.

BIAS AND FAIRNESS:

Another critical ethical consideration in AI and ML is the issue of bias and fairness in algorithmic decision-making. Machine learning models trained on biased or incomplete datasets may perpetuate and amplify existing societal biases, leading to discriminatory outcomes and unfair treatment for certain individuals or groups. Addressing algorithmic bias requires careful attention to dataset selection, feature engineering, and model evaluation, as well as the implementation of fairness-aware algorithms and bias mitigation techniques. Moreover, promoting diversity and inclusivity in AI research and development can help mitigate bias and ensure that AI systems are equitable and representative of diverse perspectives and experiences.

TRANSPARENCY AND ACCOUNTABILITY:

Ensuring transparency and accountability in AI and ML systems is essential for fostering trust and confidence among users and stakeholders. However, the complex nature of deep learning algorithms and the "black-box" nature of some AI models pose challenges to understanding how decisions are made and why. Lack of transparency can undermine accountability and raise concerns regarding algorithmic accountability and responsibility. Therefore, promoting transparency through explainable AI techniques, model documentation, and auditability becomes crucial for ensuring that AI systems are accountable for their decisions and actions.

JOB DISPLACEMENT AND ECONOMIC INEQUALITY:

The widespread adoption of AI and ML technologies has sparked fears of job displacement and economic inequality, as automation threatens to disrupt traditional employment patterns and exacerbate disparities in income and wealth. While AI has the potential to create new job opportunities and drive economic growth, its impact on the labor market remains uncertain. Addressing these challenges requires proactive measures to reskill and upskill the workforce, promote lifelong learning initiatives, and establish social safety nets to support individuals affected by technological disruptions. Moreover, ensuring equitable access to AI technologies and opportunities can help mitigate disparities and foster inclusive economic development.

ETHICAL AI GOVERNANCE AND REGULATION:

Establishing ethical AI governance frameworks and regulatory mechanisms is essential for guiding the responsible development and deployment of AI and ML technologies. However, the rapid pace of technological innovation poses challenges to traditional regulatory approaches, requiring agile and adaptive governance models that balance innovation with ethical considerations. Collaborative efforts involving government, industry, academia, and civil society are needed to develop and enforce ethical standards, codes of conduct, and regulatory guidelines that promote the responsible and ethical use of AI and ML.

In conclusion, addressing the ethical considerations surrounding AI and ML requires a holistic and multidisciplinary approach that prioritizes transparency, fairness,
accountability, and inclusivity. By proactively addressing these challenges and seizing opportunities for ethical innovation, we can harness the transformative potential of AI and ML technologies to create a more equitable, sustainable, and human-centric future for society.

CHAPTER 9

GOVERNMENT INITIATIVES: DRIVING AI AND ML ADOPTION IN INDIA

The Indian government has recognized the transformative potential of Artificial Intelligence (AI) and Machine Learning (ML) technologies and has launched several initiatives to drive their adoption and integration across various sectors. These initiatives aim to foster innovation, enhance competitiveness, and address societal challenges through the responsible and ethical use of AI and ML.

THE NATIONAL AI STRATEGY:

The Indian government has formulated a National AI Strategy to outline its vision and roadmap for AI adoption and development. This strategy encompasses initiatives such as the National AI Mission, which aims to accelerate AI research, development, and deployment across key sectors such as healthcare, agriculture, education, and governance. Additionally, the government has established AI research institutes, centers of excellence, and incubation centers to support AI innovation and entrepreneurship.

POLICY AND REGULATORY FRAMEWORKS:

The Indian government is actively working to develop policy and regulatory frameworks to govern the use of AI and ML technologies. This includes guidelines for data protection, privacy, and cybersecurity, as well as regulations to ensure ethical and responsible AI deployment. Initiatives such as the National Data Governance Framework and the Personal Data Protection Bill aim to safeguard individuals' data rights and promote trust and confidence in AI systems.

COLLABORATIVE PARTNERSHIPS:

The government is fostering collaboration between industry, academia, and research institutions to promote AI and ML innovation and entrepreneurship. Initiatives such as the AI for All program and the AI Research Consortium aim to create collaborative platforms for knowledge sharing, capacity building, and technology transfer. Moreover, the government is partnering with international organizations and leading technology companies to leverage global expertise and best practices in AI and ML.

INVESTMENT AND FUNDING:

The government is investing significant resources in AI and ML research, development, and commercialization. This includes funding programs such as the AI Startup Challenge, which provides financial support and mentorship to AI startups and entrepreneurs. Additionally, the government is incentivizing private sector investment in AI and ML through tax incentives, grants, and subsidies to accelerate technology adoption and innovation.

SKILL DEVELOPMENT AND CAPACITY BUILDING:

The government is focusing on skill

development and capacity building to ensure that India has a skilled workforce
capable of leveraging AI and ML technologies. Initiatives such as the National AI
Skilling Program and the AI Innovation Hubs aim to train professionals, students, and researchers in AI and ML concepts, tools, and techniques. Moreover, the government is promoting interdisciplinary education and collaboration to foster a holistic
understanding of AI and its applications across domains.

In conclusion, the Indian government's initiatives are playing a crucial role in driving AI and ML adoption and integration across various sectors, fostering innovation, enhancing competitiveness, and addressing societal challenges. By leveraging these initiatives and promoting collaboration between government, industry, academia, and civil society, India can harness the transformative potential of AI and ML to create a more inclusive, sustainable, and prosperous future for all.

CHAPTER 10

STARTUPS AND INNOVATION: THE ENTREPRENEURIAL SPIRIT IN AI AND ML

India's vibrant startup ecosystem is driving innovation and entrepreneurship in the field of Artificial Intelligence (AI) and Machine Learning (ML), fostering a culture of creativity, collaboration, and disruption. Startups and entrepreneurs are leveraging AI and ML technologies to address diverse challenges and opportunities across industries, from healthcare and finance to agriculture and education. In this chapter, we explore the entrepreneurial spirit in AI and ML innovation in India.

ENTREPRENEURIAL ECOSYSTEM:

India's startup ecosystem is characterized by a thriving community of entrepreneurs, investors, mentors, and accelerators, providing a conducive environment for AI and ML innovation. Startup hubs such as Bangalore, Hyderabad, and Pune are home to a burgeoning ecosystem of AI startups, research labs, and incubators, attracting talent and investment from around the world.

DIVERSE APPLICATIONS:

Startups in India are leveraging AI and ML technologies to develop innovative solutions to a wide range of challenges across sectors. In healthcare, startups are using AI for medical imaging, diagnostic decision support, and personalized medicine. In finance, AI-powered fintech startups are revolutionizing lending, investment, and insurance services. In agriculture, startups are developing precision farming
solutions, crop monitoring systems, and market forecasting tools. The diversity of applications reflects the vast potential of AI and ML to drive innovation and create value across industries.

TECH TALENT AND EXPERTISE:

India's rich pool of technical talent and expertise in computer science, engineering, and data science is fueling AI and ML innovation. Leading academic institutions such as the Indian Institutes of Technology (IITs), Indian Institutes of Science (IISc), and premier research labs are producing world-class talent in AI and ML research and development. Moreover, India's diaspora of AI professionals working in leading technology companies abroad is increasingly returning to India to contribute to the country's startup ecosystem.

ACCESS TO CAPITAL:

Startups in India have access to a diverse range of funding sources, including venture capital, angel investors, corporate accelerators, and government grants. Venture capital investment in Indian startups has been steadily increasing in recent years,
with AI and ML startups attracting significant interest from investors. Moreover, government initiatives such as the

Startup India program and the Atal Innovation
Mission provides financial support and incentives to startups to encourage innovation and entrepreneurship.

COLLABORATIVE ECOSYSTEM:

India's startup ecosystem is characterized by a spirit of collaboration and community building, with startups, investors, academia, and government working together to foster innovation and growth. Collaborative initiatives such as hackathons, accelerators, and innovation labs bring together stakeholders from diverse backgrounds to ideate, prototype, and scale AI and ML solutions. Moreover, industry partnerships and collaborations enable startups to access resources, expertise, and market opportunities to accelerate their growth and impact.

In conclusion, India's entrepreneurial spirit in AI and ML innovation is driving transformative change across industries, fueling economic growth, and

creating
societal impact. With a vibrant ecosystem
of startups, talent, and investors, India is
poised to emerge as a global leader in AI
and ML innovation, shaping the future of
technology and entrepreneurship.

CHAPTER 11

THE FUTURE WORKFORCE: ADAPTING TO THE AI AND ML ERA

The future of the workforce will undoubtedly be shaped by the continued advancement and integration of artificial intelligence (AI) and machine learning (ML) technologies. As these technologies become more sophisticated and pervasive across industries, workers will need to adapt to new ways of working and acquire new skills to remain relevant and competitive in the job market. Here are some key considerations for adapting to the AI and ML era:

UPSKILLING AND RESKILLING

With the automation of routine tasks, many traditional jobs may become obsolete. Workers will need to upskill or reskill to take on roles that require human skills such as creativity, critical thinking, emotional intelligence, and complex problem-solving.

Training programs and educational initiatives will be crucial in helping workers transition into these new roles.

EMBRACING AI AS A TOOL

Rather than seeing AI and ML as threats to job security, workers should embrace them as tools to enhance their productivity and decision-making abilities.
Understanding how to leverage AI and ML algorithms to streamline processes, analyze data, and generate insights will be essential across various industries.

ADOPTING A GROWTH MINDSET

The rapid pace of technological advancement means that workers will need to adopt a growth mindset and be open to continuous learning and adaptation. Flexibility, curiosity, and a willingness to experiment with new technologies and ways of
working will be valuable traits in the AI and ML era.

COLLABORATION WITH MACHINES

AI and ML technologies will augment human capabilities rather than replace them entirely. Workers will need to learn how to collaborate effectively with intelligent machines, leveraging their strengths to achieve better outcomes. This might involve tasks such as data labeling, model training, and interpreting AI-generated insights.

ETHICAL AND RESPONSIBLE AI USE

As AI and ML technologies become more powerful, there will be increasing concerns about their ethical use, including issues related to bias, privacy, and accountability. Workers will need to be educated about these ethical considerations and advocate for responsible AI practices within their organizations.

ADAPTING TO REMOTE WORK AND DIGITAL COLLABORATION

The COVID-19 pandemic has accelerated the shift towards remote work and digital collaboration, a trend that is likely to continue in the AI and ML era. Workers will need to adapt to virtual work environments, hone their digital communication skills, and find new ways to collaborate effectively with colleagues and AI-powered tools.

ENTREPRENEURIAL MINDSET

In addition to traditional employment opportunities, the AI and ML era will also create new possibilities for entrepreneurship and innovation. Workers with an entrepreneurial mindset may seize opportunities to create new businesses, products, or services that leverage AI and ML technologies to address emerging needs and
challenges.

In summary, adapting to the AI and ML era will require a combination of upskilling, embracing technological change, fostering a growth mindset, collaborating with intelligent machines, addressing ethical considerations, adapting to remote work, and cultivating an entrepreneurial spirit. By proactively embracing these changes, workers can position themselves for success in the future workforce.

CHAPTER 12

DATA SECURITY AND PRIVACY IN THE AGE OF AI AND ML

In the rapidly evolving landscape of artificial intelligence (AI) and machine learning (ML), data security and privacy have become paramount concerns. As these technologies increasingly rely on vast amounts of data to train models and make predictions, safeguarding sensitive information and ensuring user privacy have become significant challenges. This chapter explores the unique considerations and strategies for addressing data security and privacy in the age of AI and ML.

DATA ENCRYPTION AND ACCESS CONTROL

To protect data from unauthorized access, encryption techniques such as secure socket layer (SSL) encryption and advanced encryption standard (AES) can be employed. Additionally, implementing strict access controls based on roles and permissions ensures that only authorized individuals can access sensitive data.

ANONYMIZATION AND PSEUDONYMIZATION

Anonymizing or pseudonymizing data before it is used for training AI and ML models can help protect individual privacy. By removing personally identifiable information (PII) or substituting it with artificial identifiers, organizations can minimize the risk of data breaches and unauthorized disclosures.

SECURE DATA SHARING PROTOCOLS

When sharing data with third parties or collaborating with external partners, employing secure data sharing protocols such as secure multiparty computation (SMC) or federated learning can help maintain data security and privacy. These protocols enable organizations to collaborate on model training without sharing raw data.

ADVERSARIAL ROBUSTNESS

AI and ML models are vulnerable to adversarial attacks, where malicious actors manipulate input data to deceive the model's predictions. Implementing techniques such as adversarial training and robust model validation can enhance the resilience of AI and ML systems against such attacks.

PRIVACY-PRESERVING AI AND ML TECHNIQUES

Privacy-preserving AI and ML techniques, such as differential privacy and homomorphic encryption, enable organizations to derive insights from sensitive data without compromising individual privacy. These techniques add noise or encrypt data in a way that preserves privacy while still allowing for meaningful analysis.

REGULATORY COMPLIANCE

Compliance with data protection regulations such as the General Data Protection Regulation (GDPR) and the California

Consumer Privacy Act (CCPA) is essential for ensuring data security and privacy in the age of AI and ML. Organizations must understand their obligations under these regulations and implement measures to comply with data protection requirements.

ETHICAL CONSIDERATIONS

Beyond regulatory compliance, organizations must consider the ethical implications of their AI and ML initiatives, particularly regarding data security and privacy. Adopting ethical frameworks such as fairness, accountability, and transparency (FAT) principles can help guide decision-making and mitigate potential harms associated with AI and ML technologies.

CONTINUOUS MONITORING AND AUDITING

Establishing robust monitoring and auditing mechanisms is crucial for detecting and addressing potential security and privacy breaches in AI and ML systems. Regularly reviewing access logs, conducting security assessments, and performing penetration testing help ensure the ongoing integrity of data security and privacy measures.

In conclusion, data security and privacy are central concerns in the age of AI and ML, requiring organizations to implement comprehensive strategies to protect sensitive information and uphold user privacy rights. By leveraging encryption, access controls, privacy-preserving techniques, regulatory compliance, ethical considerations, and continuous monitoring, organizations can mitigate the risks associated with AI and ML while fostering trust and accountability in their data practices

CHAPTER 13

AI AND ML IN LEGAL SYSTEMS: NAVIGATING REGULATORY LANDSCAPES

The integration of artificial intelligence (AI) and machine learning (ML) technologies into legal systems presents both opportunities and challenges. As these technologies become increasingly sophisticated, legal professionals must navigate complex regulatory landscapes to ensure compliance, protect individual rights, and uphold the principles of justice. This chapter explores the role of AI and ML in legal systems and provides guidance on navigating regulatory frameworks.

LEGAL RESEARCH AND DOCUMENT ANALYSIS

AI and ML algorithms can streamline legal research and document analysis, enabling legal professionals to access relevant case law, statutes, and legal documents more efficiently. However, the use of AI and ML in legal research raises questions about the accuracy, reliability, and bias of algorithmic outputs, requiring careful scrutiny and validation by human experts.

PREDICTIVE ANALYTICS IN LEGAL DECISION-MAKING

Predictive analytics powered by ML algorithms have the potential to assist judges, lawyers, and policymakers in making informed decisions on case outcomes, sentencing, and legal strategy. However, concerns about fairness, transparency, and accountability arise when using predictive models in legal decision-making, necessitating the development of ethical guidelines and regulatory oversight.

CONTRACT REVIEW AND DUE DILIGENCE

AI and ML technologies can automate contract review and due diligence processes, identifying relevant clauses, potential risks, and discrepancies more quickly and accurately than manual methods. Legal professionals must ensure the accuracy and completeness of AI-generated insights while addressing concerns about data privacy, confidentiality, and regulatory compliance.

LEGAL COMPLIANCE AND REGULATORY MONITORING

AI and ML systems can assist organizations in monitoring regulatory compliance, detecting potential violations, and mitigating legal risks. However, the use of AI in regulatory compliance raises challenges related to data privacy, algorithmic transparency, and regulatory interpretation, requiring close collaboration between legal and technical experts.

ETHICAL AND PROFESSIONAL RESPONSIBILITIES

Legal professionals have ethical and professional responsibilities to uphold legal standards, protect client confidentiality, and ensure the fair and impartial administration of justice. When using AI and ML technologies in legal practice, lawyers must adhere to ethical guidelines, disclose the limitations of algorithmic tools, and exercise independent professional judgment.

DATA PRIVACY AND SECURITY

Protecting the privacy and security of sensitive legal data is essential to maintaining client trust and complying with legal and regulatory requirements. Legal professionals must implement robust data protection measures, including encryption, access controls, and data anonymization, to safeguard confidential information from unauthorized access or disclosure.

REGULATORY COMPLIANCE AND CERTIFICATION

Regulatory bodies and professional associations play a crucial role in establishing standards, guidelines, and certification programs for the responsible use of AI and ML in legal practice. Legal professionals must stay informed about relevant regulations, undergo training on AI ethics and compliance, and obtain certifications to demonstrate competence in using AI and ML technologies.

CONTINUOUS MONITORING AND EVALUATION

Legal organizations should regularly monitor and evaluate the performance, reliability, and ethical implications of AI and ML systems deployed in legal practice. Conducting audits, soliciting feedback from stakeholders, and conducting impact assessments help identify and address potential biases, errors, or unintended consequences of algorithmic decision-making.

The integration of AI and ML technologies into legal systems offers opportunities to improve efficiency, accuracy, and access to justice. However, legal professionals must navigate complex regulatory landscapes, address ethical considerations, and uphold professional standards to ensure the responsible and ethical use of AI and ML in legal practice. By staying informed, collaborating across disciplines, and prioritizing transparency and accountability, legal professionals can harness the transformative potential of AI and ML while upholding the principles of justice and fairness.

CHAPTER 14

BEYOND BORDERS: INDIA'S ROLE IN GLOBAL AI AND ML ADVANCEMENTS

India has emerged as a significant player in the global landscape of artificial intelligence (AI) and machine learning (ML) advancements. With its large pool of highly skilled tech talent, thriving startup ecosystem, and government initiatives to promote innovation and digital transformation, India is poised to make significant contributions to the development and adoption of AI and ML technologies worldwide. This chapter explores India's role in global AI and ML advancements and examines the factors driving its growth and impact.

TALENT POOL AND EDUCATION

India boasts a vast talent pool of engineers, data scientists, and AI researchers, fueled by its extensive network of technical universities, research institutions, and specialized training programs. Indian tech professionals are sought after globally for their expertise in AI and ML, contributing to the development of cutting-edge technologies and solutions.

STARTUP ECOSYSTEM

India's vibrant startup ecosystem is a hotbed of innovation in AI and ML, with numerous startups leveraging these technologies to address diverse challenges across industries such as healthcare, finance, agriculture, and e-commerce. The availability of venture capital funding, supportive government policies, and access to mentorship and networking opportunities have fueled the growth of AI and ML startups in India.

INDUSTRY ADOPTION AND INNOVATION

Indian enterprises are increasingly embracing AI and ML to drive efficiency, innovation, and competitive advantage.

Companies across sectors are deploying AI-powered solutions for predictive analytics, customer engagement, process automation, and personalized recommendations, transforming how businesses operate and deliver value to customers.

GOVERNMENT INITIATIVES

The Indian government has launched several initiatives to foster the growth of AI and ML in the country, including the National AI Strategy, the AI for All Initiative, and the National Program on AI. These initiatives aim to promote research and development, skill development, industry collaboration, and the ethical adoption of AI and ML technologies across sectors.

INTERNATIONAL COLLABORATION

India actively collaborates with international partners, research institutions, and tech companies to advance AI and ML research, innovation, and standards. Partnerships with countries such as the United States,

Canada, Israel, and Singapore facilitate knowledge exchange, joint research projects, and talent mobility, driving global AI and ML advancements.

ETHICAL AND INCLUSIVE AI

India is committed to developing AI and ML technologies that are ethical, inclusive, and beneficial for society. Initiatives such as the Responsible AI for Youth program, the National AI Ethics Committee, and the AI for Social Good Challenge focus on addressing ethical considerations, mitigating biases, and leveraging AI for social impact and sustainable development.

CHALLENGES AND OPPORTUNITIES

Despite its significant progress, India faces challenges such as inadequate infrastructure, digital divide, data privacy concerns, and talent retention. Addressing these challenges requires concerted efforts from government, industry, academia, and civil society to build robust digital infrastructure, promote data privacy and security, and ensure equitable access to AI and ML technologies.

FUTURE OUTLOOK

India's role in global AI and ML advancements is poised to expand further in the coming years, driven by ongoing investments in research and development, talent

development, industry innovation, and international collaboration. With its diverse talent pool, entrepreneurial spirit, and commitment to ethical and inclusive AI, India is well-positioned to shape the future of AI and ML on the global stage.

In conclusion, India's growing influence in the field of AI and ML reflects its commitment to innovation, talent development, and collaboration on the global stage. By leveraging its strengths in talent, technology, and entrepreneurship, India has the potential to drive significant advancements in AI and ML that benefit both the country and the world at large.

CHAPTER 15
THE ROAD AHEAD: ANTICIPATING THE NEXT WAVE OF AI AND ML DEVELOPMENT

As artificial intelligence (AI) and machine learning (ML) continue to evolve at a rapid pace, it is essential to anticipate the next wave of development in these fields. This chapter explores emerging trends, challenges, and opportunities shaping the future of AI and ML and provides insights into the road ahead.

EXPLAINABLE AI

There is a growing need for AI systems to be explainable and transparent, enabling users to understand how decisions are made and to trust the outcomes. Research in explainable AI aims to develop algorithms and techniques that provide interpretable explanations for AI predictions and recommendations, fostering trust and accountability.

INTERDISCIPLINARY RESEARCH

The future of AI and ML development will be characterized by interdisciplinary collaboration, bringing together experts from diverse fields such as computer science, neuroscience, psychology, sociology, and ethics. By integrating insights from multiple disciplines, researchers can address complex challenges and develop AI systems that are more human-centered and socially responsible.

AI FOR HEALTHCARE

AI and ML have the potential to revolutionize healthcare by enabling personalized treatment plans, early disease detection, medical imaging analysis, drug discovery, and virtual health assistants. The integration of AI into healthcare systems will improve patient outcomes, reduce costs, and enhance the efficiency of healthcare delivery.

EDGE COMPUTING AND IOT

The proliferation of Internet of Things (IoT) devices and the advent of edge computing are driving demand for AI and ML solutions that can operate efficiently at the network edge. Edge AI enables real-time processing and analysis of data locally on devices, reducing latency, conserving bandwidth, and enhancing privacy and security.

RESPONSIBLE AI GOVERNANCE

As AI technologies become more pervasive, there is a pressing need for frameworks and regulations that govern their ethical use, accountability, and societal impact. Governments, industry stakeholders, and civil society organizations are working to develop guidelines, standards, and policies for responsible AI governance, addressing issues such as bias, fairness, privacy, and algorithmic transparency.

CONTINUAL LEARNING AND ADAPTATION

AI systems that can continually learn and adapt to changing environments, preferences, and user feedback will be essential for addressing dynamic and evolving challenges. Research in lifelong learning, meta-learning, and reinforcement learning aims to develop AI models that can acquire new knowledge, skills, & capabilities over time, improving their performance and robustness.

AI ETHICS AND BIAS MITIGATION

Mitigating biases and ensuring fairness, equity, & inclusivity in AI systems are critical challenges that require concerted efforts from researchers, developers, policymakers, & stakeholders. Techniques such as fairness-aware learning, bias detection & mitigation, and diversity-aware AI aim to address biases in data, algorithms & decision-making processes, promoting ethical and equitable AI deployment.

HUMAN-AI COLLABORATION

The future of AI and ML development will involve closer collaboration between humans and machines, leveraging the complementary strengths of both. Research in human-AI interaction, human-centered AI design, and collaborative AI aims to create AI systems that augment human intelligence, enhance productivity, and empower users to achieve their goals more effectively.

CONCLUSION

In conclusion, the next wave of AI and ML development will be characterized by advances in explainable AI, interdisciplinary research, healthcare applications, edge computing, responsible AI governance, continual learning, bias mitigation, and human-AI collaboration. By anticipating emerging trends and addressing key challenges, researchers, practitioners, and policymakers can shape a future where AI and ML technologies benefit society while upholding ethical principles and human values.

In "Revolutionizing India: The Impact of Artificial Intelligence & Machine Learning," we have embarked on a journey to explore the transformative potential of AI and ML in shaping India's future. Through the lens of technological innovation, talent development, and societal impact, we have witnessed the profound changes unfolding across various sectors, from healthcare and agriculture to finance and governance.

As India embraces the fourth industrial revolution, AI and ML are poised to revolutionize how we live, work, and interact with the world around us. The convergence of data, algorithms, and computing power has unlocked new opportunities for innovation, efficiency, and inclusivity, empowering individuals, businesses, and governments to address complex challenges and seize new possibilities.

From empowering farmers with precision agriculture techniques to revolutionizing healthcare delivery through predictive analytics and telemedicine, AI and ML are driving tangible improvements in people's lives, enhancing productivity, and driving economic growth.

However, as we chart this path of progress, we must also navigate challenges such as ethical considerations, data privacy concerns, and the equitable distribution of benefits. As stewards of this technological revolution, we must ensure that AI and ML are deployed responsibly, ethically, and inclusively, with a focus on maximizing societal benefit and minimizing harm. Looking ahead, the future of India lies in harnessing the transformative potential of AI and ML to build a more prosperous, equitable, and sustainable society. By

fostering innovation, investing in talent development, and embracing a culture of collaboration and inclusivity, we can unlock new frontiers of possibility and shape a brighter future for generations to come. "Revolutionizing India: The Impact of Artificial Intelligence & Machine Learning" serves as a testament to the power of technology to drive positive change and inspire collective action towards a better tomorrow. As we continue on this journey of discovery and innovation, let us remain committed to leveraging AI and ML for the greater good, empowering individuals and communities to thrive in the digital age.

"In the infusion of AI and ML into human life, we don't just witness technological advancement; we witness the evolution of possibility, where innovation meets the human spirit to redefine what it means to thrive in a connected world."

Akhilesh Shukla